Enjoy Physics Through Projects
-A Play With Diodes

Enjoy Physics Through Projects -
-A Play With Diodes

Enjoy Physics Through Projects
-A Play With Diodes

S. Sankararaman

ADHYAYAN
ADHYAYAN PUBLISHERS & DISTRIBUTORS
NEW DELHI (INDIA)

Published by
ADHYAYAN PUBLISHERS & DISTRIBUTORS
4378/4B, 105, J.M.D. House, Murari Lal Street
Ansari Road Darya Ganj, New Delhi - 110002
Ph.: 011-23263018, 011-23277156, Fax: 011-23280028
Email: adhyayanpublishers@yahoo.com

Enjoy Physics Through Projects -A Play With Diodes

Edition 2011
ISBN 978-81-8435-256-6

Printed in India

Published by Harish Chandra Yadav for Adhyayan Publishers & Distributors, Laser Typeseting at A.V. Rao and Printed at Tarun Offset Printers, Delhi

PREFACE

Physics is a fascinating subject that can be enjoyed if properly understood. Most of the students do not know how to enjoy Physics. One way to enjoy the subject is to keep your eyes open to the surroundings and study the theories well. You will see Physics as an omnipresent reality. Then keep on asking what, why and how. Think for the answer. Read books. Trouble your teachers with doubts. Try to relate the theories you have studied in the classroom to the real world. I am sure the door of the wonderful world of Physics will be opened before you.

Experimentation is an unavoidable part of science as they enhance the level of understanding and give a vivid picture of the theory. It can take you to the world of new ideas, problems and discoveries. To promote originality in thinking and to develop capabilities to view things in a scientific way project works are included as part of curriculum. Nowadays this has become a problem for students. They do not know what and how to do a project work. I have seen students wandering here and there in search of a project work. When they fail in finding a Physics project they turn to some hobby electronic circuits, as project reports are compulsory in degree and post graduate courses.

In this book "Enjoy Physics through projects - A play with diodes" I have made a humble effort to enable you to think and to find a good project work. If you focus your thoughts on a specific physical problem, work on it and prepare a report of that, be sure, it will be an excellent project work. In other words, project work is a planned programme of work on a problem. In the last section, details on how to prepare a project report is also given so as to enable the students to prepare their project report.

I am sure that each study will teach you something. One thing which you have to keep in mind is that you should never make a foolish attempt of classifying a topic as that of plus two (intermediate) level, under-graduate level or post-graduate level. Remember, there is one and only one Physics for all classes and we explore it from the level of our understanding. If a person classifies a subject as of lower level or higher level one can be sure that his level of Physics in him is very low. The higher our level of understanding the greater is the information we deduce from small things.

As an example, from the simple V-I characteristics of a pn junction diode how to develop a project work is illustrated through eleven projects described in this book. I have done all these projects and any one can carry out these projects within the limited facilities of colleges. While discussing the projects, in some places, I have made certain approximations without affecting your understanding of the subject. Though the limitations of available measuring devices can lower the level of accuracy of the measured value it will never lower the level of information you could get from the project.

My humble request is that you need not unnecessarily worry over the lack of accuracy due to the unavailability of measuring instruments. What counts is the idea. Try to develop a skill of innovative thinking. If we examine anything there will be something interesting to study. Try to find that. My appeal to students is that we should join hands to achieve this goal. Carry out the projects given in this book and make your own modifications or extensions. Do not forget to study relevant theories also. I am sure that you will start enjoying Physics by the time I come with the second volume of Enjoy Physics through Projects.

Dr. S. Sankararaman

CONTENTS

APPENDIX

1

A TEMPERATURE SENSOR

We are all aware of the significant role of semiconductor diodes in Physics and Electronics. The property of semiconductor,viz., the negative temperature coefficient of resistance makes diode suitable for a variety of applications. The volt (V)- ampere (I) characteristics of the diode, though seems to be simple to carry out the experiment, it is a treasure of information regarding the diode. Most of the students study the V–I characteristics of diode without knowing its significance.

The change of biasing voltage (forward or reverse) changes the width of the depletion region and the current through the diode. We know that the diode current we measure include contributions from generation recombination current and diffusion current. As the rise in temperature is capable of bringing more and more electrons from the valance band to the conduction band of the semiconductor material of the diode, the current through pn junction changes. All these information are contained in the diode equation

$$I = I_0\left[\exp\left(\frac{eV}{\eta kT}\right) - 1\right]$$

Neglecting one we get,

$$I \cong I_0\left[\exp\left(\frac{eV}{\eta kT}\right)\right] \qquad (1.1)$$

Where I is the diode current or the current through the forward biased pn junction at temperature T (in Kelvin scale) and biasing voltage V. I_0, e, k and h denote the reverse saturation current, electronic charge, Boltzmann's constant and ideality factor respectively.

From the diode equation we can see that the diode current depends on absolute temperature. As a simple project students can undertake the work of studying how the diode current varies with temperature for a given forward biasing voltage. Once this relation is studied and represented graphically, try to measure the unknown temperature with the help of the graph already plotted. Further the experiment can also be carried out with different diodes. Do you find any change? If so start asking why and how? I am sure that you will end up with a good project work.

The discussion of detailed theory of the experiment is beyond the scope of this book. Hence I will be giving you only the essentials in a most concise manner. The present work can be carried out with the help of the circuit (Fig. 1) given below.

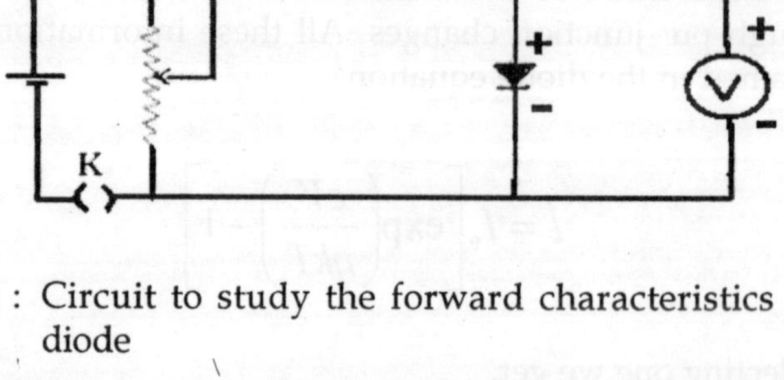

Fig.1 : Circuit to study the forward characteristics of a diode

For doing this project work we need a 2 V battery, rheostat, key, voltmeter, an ammeter and a diode (say IN 4007). If possible it is better to use a good digital multimeter so that we can easily measure even a slight change in the diode current.

Having the connections done, the biasing voltage can be gradually increased by changing the sliding contact of the rheostat. The ammeter reading, which indicates the current through the diode, is noted for each forward biasing voltage. Then a graph is plotted with the biasing voltage V along the x–axis and the current through the diode along the y- axis. The typical V-I characteristic curves for a silicon, germanium and gallium arsenide diodes are shown in fig. 2.

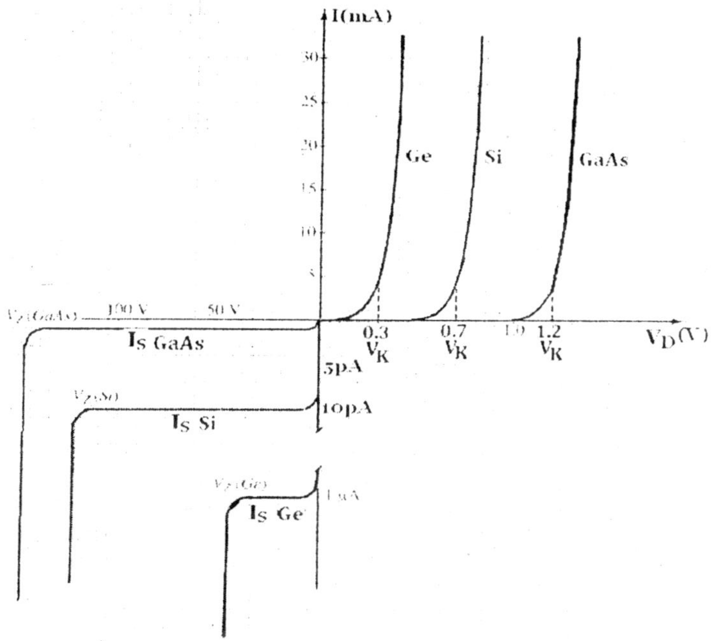

Fig.2 : V-I characteristic curve fòr a silicon, germanium and gallium arsenide diode

From the V-I characteristics the knee voltage (the forward biasing voltage at which the current increases rapidly) can be found out. To use the diode as a temperature sensor it can be operated either by keeping the biasing voltage a constant or by keeping the forward current a constant. From fig. 3 we can see that for a given forward current I, V decreases. In other words for a given voltage V, I increases. The shifting of the characteristic curve with increase and decrease in temperature is given in the table.

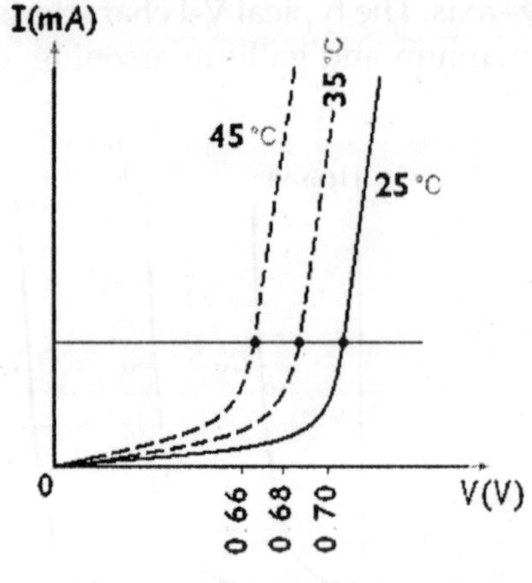

Fig. 3. V-I forward characteristics at different temperatures for a silicon diode

Temperature	Forward bias region	Reverse bias region
Increase	Shift **Left** at rate of 2.5 mV per ^{0}C	Reverse Sat. Current doubles per 10^{0}C
Decrease	Shift **Right** at rate of of 2.5 mV per ^{0}C	Reverse Sat. Current doubles per 10^{0}C

If you want to keep the current constant, it is better to operate the diode in the linear portion of the V-I characteristics. Then only you will be able to get a noticeable change in V. While doing like this, you need change the rheostat before taking the reading at each temperature, for keeping I constant.

In the second method one can keep the applied biasing voltage a constant. If you want to keep the voltage constant, it is better to operate the diode below the knee voltage. Here as the applied voltage is not changing during the experiment one can easily note the changes in the forward current.

The second method is easy compared to the first as frequent adjustments in the rheostat are not needed. Since in the first method we have to record the voltage at each temperature it will be difficult to watch the thermometer, voltmeter and ammeter readings simultaneously while adjusting the rheostat. The error in the reading can be avoided if we have a constant temperature bath so that it can be set at any desired temperature. Normally we use an oil bath or water bath for heating the diode. It does not have any provision for keeping the temperature constant. Under such a condition it is good to follow the second method.

In both cases we make use of the principle of temperature dependence of carrier transport. We know that the diode current comprises recombination current and diffusion current. The diffusion current has temperature dependence. Under very low bias or temperature the recombination current dominates. With the increase of temperature or biasing voltage the diffusion current increases. When the temperature of a semiconductor material is increased the additional thermal energy creates more and more free charge carriers. That is, more electron - hole pairs are generated. Thus, the forward current also increases. The effect of increase of current in a forward biased pn junction is shown in fig 3. It is found that the temperature coefficient for the forward voltage of pn junction is approximately -2.5 mV/°C for silicon or germanium at room temperature.

To carry out the experiment, first study the V–I characteristics of the diode at room temperature. Then the temperature of the diode is varied by heating it by keeping in an oil bath or pure water bath. The forward current is noted at each temperature keeping the biasing voltage constant. The experiment can be repeated for different biasing voltages. Now a graph is plotted with temperature T along the x–axis and current I along the y – axis. We get a curve corresponding to the exponential relation given by equation 1.1

Taking logarithm on either sides of equation 1 we get

$$\ln I = \ln I_0 + \left(\frac{eV}{\eta k}\right)\frac{1}{T} \tag{1.2}$$

If ($\ln I$) – ($1/T$) graph is plotted we get a straight line as V is kept constant. Thus for each value of V we get a straight line graph. These graphs can be called calibration graphs as they can be used for the measurement of unknown temperature.

Having the diode characteristics calibrated against standard thermometer reading, it can be now used to measure unknown temperature, say, the melting point of wax. For this the wax is taken in a test tube and heated in a water bath. The forward biased calibrated diode also is kept in the water bath. When the wax melts the diode current is noted. From the calibration graph already plotted the melting point of wax can be found. Similarly the diode can be used to measure any temperature. Note that we have to apply the same biasing voltage for calibration and temperature measurement.

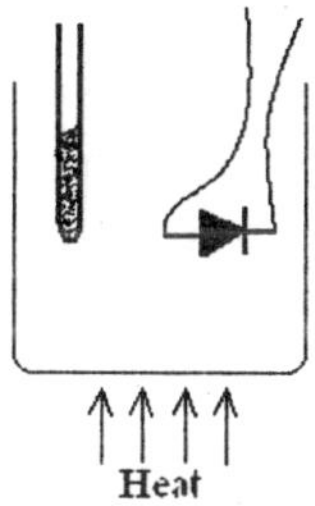

OBSERVATIONS

(i) V – I characteristics at room temperature.

Voltage (*V*)	Current (*mA*)

(ii) Variation of diode current with temperature T

(a) V= ____V (b) V= ____V (c) V= ____V

Temperature *T (K)*	Current *I* *(mA)*	ln *I*	1/*T*

(iii) Plotting of graph

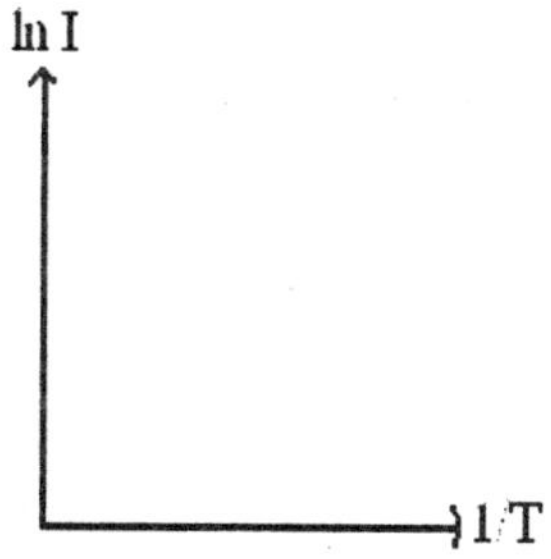

OR

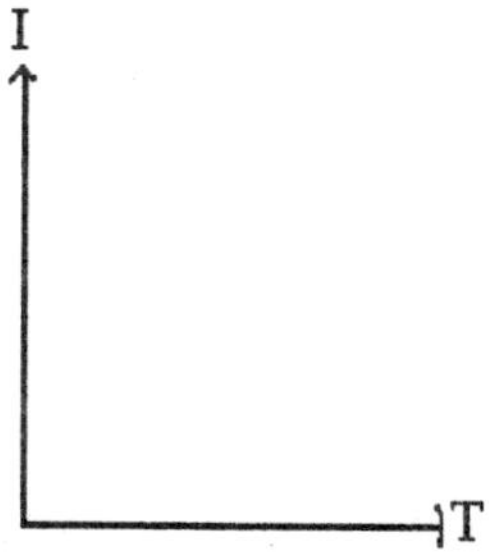

The accuracy of temperature measurement depends on the accuracy of calibration. This in turn depends on the accurate measurement of current or voltage. Hence sensitive digital multimeters are to be used in the experiment for current or voltage measurement.

2
VARIATION OF IDEALITY FACTOR

The current through a pn junction is the sum of recombination current and diffusion current. From the discussion of theory of forward biased pn junction diode we can see that the current depends on the carrier diffusion coefficient, doping concentration, width of depletion layer,etc., all of which are interrelated. A study of ideality factor can tell us about the nature of current through the diode.

Thus the ideality factor denotes the type of current through the pn junction. That is, whether recombination current or diffusion current dominates or both are comparable. At low current density the recombination current component domi nates and at higher current density the diffusion current dominates.

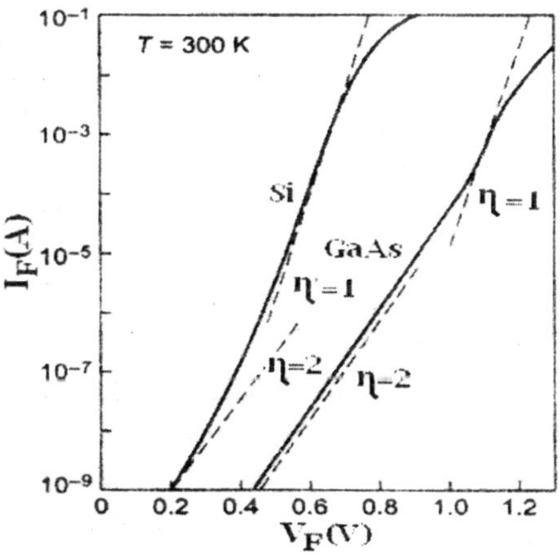

Fig.1 : Forward characteristics of silicon and GaAs diode.

Accordingly depending upon the nature of current, the value of ideality factor lies between one and two. Under low forward bias as the recombination current dominates the value of ideality factor can be taken as two and for large forward bias voltage as the diffusion current dominates the value of ideality factor can be taken as one. This is the reason why we take the value of ideality factor as two for silicon under low forward bias. The change in the ideality factor value is shown in fig.1. From the figure it is evident that there is a transition region in the V-I characteristics where the value of ideality factor lies between two and one. In this region the contribution of diffusion current and recombination current to the total current is comparable or significant.

While doing the experiment for the determination of bandgap energy of silicon or germanium using diodes, students often do certain manipulations, in the graph they draw, for getting the correct value. Here the problem arises because students have blindly taken the value of ideality factor for silicon as two and germanium as one. They never think about the region of V-I characteristics where the value of ideality factor is two and where it is one. The value of ideality factor becomes a problem in silicon or GaAs diodes as the diode current has contributions from both currents. The change in the value of ideality factor for silicon and GaAs diode is shown in fig. 1. From the graph shown in fig. 1 it is clear that, in the case of silicon diode, only under very low forward bias we can take the value of ideality factor as two.

Hence, prior to the experiment if the students have made any effort to find the value of ideality factor for the diode under the operating condition (biasing voltage or forward current through the diode) they need not have to go for manipulation. With this value of ideality factor if they try to calculate the bandgap energy they will get more correct answer. Therefore the study of ideality factor is important.

Here evolves the scope of a project work. Students should try to find out the value of ideality factor for a given diode using diode equation. The project can also be extended to find out whether there is any change in the value of ideality factor in different types of silicon diodes like IN 4001 - IN 4007, IN 5402, IN 5408, BY159, etc. We know that when the diffusion current dominates the value of ideality factor is one. We also know the change in the diffusion of carriers across the junction due to change in temperature. Therefore it will be good to study, how the ideality factor changes with temperature. Once we investigate such things we can say that we have done a good project work.

For carrying out the experiment we can use the same circuit given as fig. 1 of project 1. From the diode equation we can write

$$\ln I = \ln I_0 + \frac{1}{\eta}\left(\frac{eV}{kT}\right) \tag{2.1}$$

From the V-I characteristic study we can tabulate the current through the pn junction for different biasing voltages at a given temperature. If we plot *ln I* against *V* we get the slope of the graph as $\frac{e}{\eta kT}$.

i.e. $$\text{slope} = \frac{e}{\eta kT}$$

Since $$\frac{e}{k} = 11600, \quad \eta = \frac{11600}{slope.T} \tag{2.2}$$

where the temperature T is in Kelvin scale.

The experiment can be repeated with different types of diode that are commonly available in electronic shops. We can see that the value of ideality factor is less than two in most cases. It is advisable to study the V-I characteristics at low biasing voltage. Figure 1 can tell you the biasing voltage required to get the value of ideality factor as two, between one and two and one for silicon diode. As we increase the biasing voltage, the contribution of diffusion current to the total forward current also increases and the value of ideality factor decreases. The value of ideality factor will be two if the contribution of recombination current alone is considered or when it dominates.

The experiment can also be carried out at different temperatures keeping the biasing voltage constant. This work can tell you the dependence of diffusion current on temperature. As temperature increases the contribution of diffusion current also increases. Now plot lnI vs V graph and find its slope. The value of ideality factor at various temperatures can be calculated using equation 2.2. Could you find any change? If yes be happy otherwise don't worry. I am sure that you will never get the value of ideality factor as two at higher temperature for silicon. The ideal diffusion, recombination and total current in a forward biased pn junction is shown in fig. 2

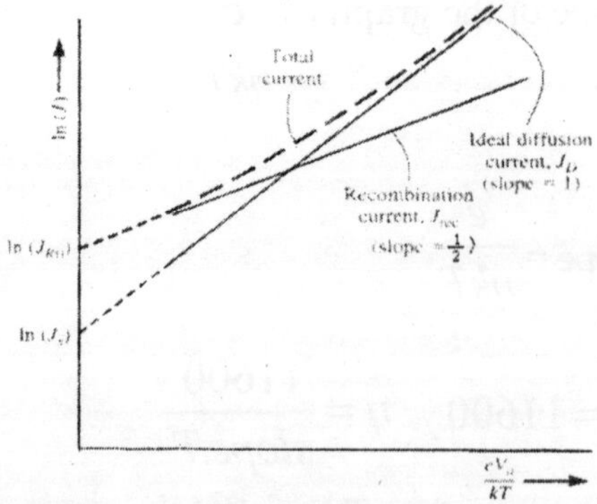

Fig.2 : Ideal diffusion, recombination and total current in a forward biased pn junction.

3
STUDY OF DIFFUSION CAPACITANCE

We have studied in the theory of semiconductor diode about the formation of a depletion region at the pn junction. The width of the depletion layer changes with the biasing voltage. In the depletion layer we have positive immobile ions on the n side and the negative immobile ions on the p side. These immobile ions are responsible for the junction capacitance or depletion capacitance.

When pn junction is forward biased, the depletion capacitance of the junction increases due to the decrease in the width of depletion layer. In addition, due to the motion and storage of minority carriers on either side of the junction, a capacitance known as the storage or diffusion capacitance (C_d) is introduced. The storage or diffusion capacitance takes into account of the capcitive effects of the carrier injected into each side of the junction when it is forward biased. The introduction of the diffusion capacitance leads to the restriction of the diode at high frequency operations. The diffusion capacitance is important at low frequencies and forward biased conditions.

In a forward biased pn junction diode electron density and hole density co exist in the neutral *n* region. When the forward bias changes due to the ac component voltage the minority carrier concentration also changes. This charge dQ is alternately being charged and discharged through the junction as the voltage across the junction changes. This change in stored minority carriers as a function of the change in voltage is the diffusion capacitance.

If the change in the number of holes stored per unit area of the n-layer is dQ_p, when the applied forward bias changes by dV, the diffusion capacitance due to the stored holes on the n-side is

$$C_p = \frac{dQ_p}{dV} \tag{3.1}$$

The diffusion (storage) capacitance per unit area due to electrons on the p side is

$$C_n = \frac{dQ_n}{dV} \tag{3.2}$$

In a forward biased diode the two capacitors C_p and C_n behave as if they are connected in parallel to each other. Hence the total diffusion capacitance per unit area of the junction is

$$C_d = C_p + C_n \tag{3.3}$$

The simplest method for the determination of diffusion capacitance is to use the principle of formation of Lissajous figure on a CRO (cathode ray oscilloscope) using the circuit given in fig. 1.

A 6-0-6 or 12-0-12 transformer can be used in the circuit. In order to change the voltage applied across the diode (*D*) and variable resistor (*R*), a rheostat can be used in voltage divider configuration. The principle behind the circuit is that there is a phase difference between the voltage across the diode and the resistor. The diode used here plays the role of a capacitor. Thus the diode-resistor series connection can be considered as equivalent to a capacitor–resistor series circuit. It is the junction capacitance that makes the diode to act like a capacitor. We make use of this diffusion capacitance for getting the Lissajous pattern.

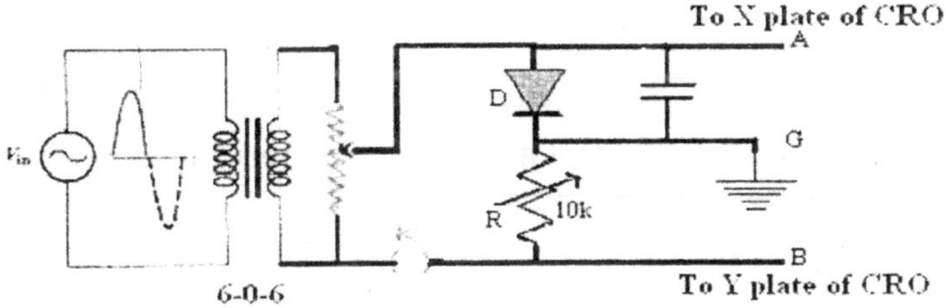

Fig.1 : Experimental circuit diagram.

The voltage across the diode and the resistor are fed to the CRO to the *X* and *Y* plates respectively, i.e. positive probe of channel one to the point *A* in the circuit and that of channel two to the point *B*. The negative probe of both channels are connected to the point *G*. Now operate the CRO in the *XY* mode. By suitably adjusting the rheostat a fixed voltage is applied to the diode and the variable resistor. By varying the variable resistor the voltage across the diode can be adjusted. The horizontal and vertical gain controls of the CRO are adjusted to get the diode characteristics pattern on the screen. You may get any one of the four patterns shown below. All are the same. One can invert the pattern by pressing the inverting button INV on the CRO. One can also interchange the connections to the *X* and *Y* plate. Trace this pattern and draw it on a graph paper. The splitting of the horizontal arm of the graph is due to the diffusion capacitance (C_d) and the vertical arm is due to the transition capacitance (C_t).

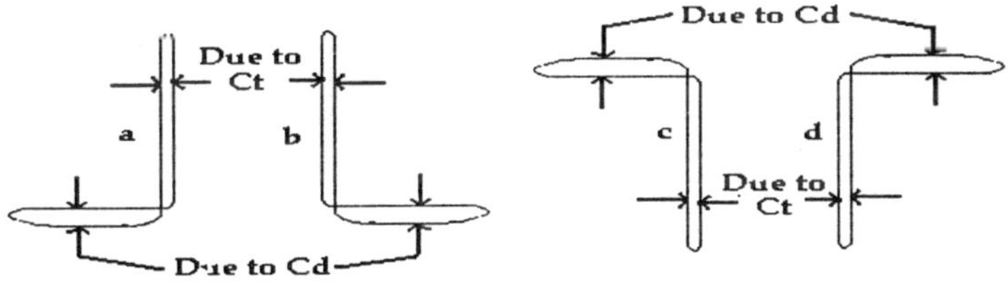

Fig.2 : The possible Lissajous pattern on the CRO

We know that the effective capacitance of two parallel capacitors is the sum of their capacitances. Now connect capacitors of suitable values across the diode and keep watch the change in the width of the split trace. It can be seen that the width of the horizontal arm of the trace increases as we connect a capacitor. When the capacitance value of the capacitor thus connected becomes equal to the diffusion capacitance the width gets doubled as shown in fig. 3. This is because the effective capacitance gets doubled. From this we can find the value of the diffusion capacitance. The experiment is repeated with capacitors of slightly different capacitances. Trace the pattern and note the width of the trace in each case. From these observations the diffusion capacitance can be calculated using the equation 3.4.

Capacitance (Micro farad)	Width of split trace
C_d	$X_1 =$
$C_d + C$	$X_2 =$

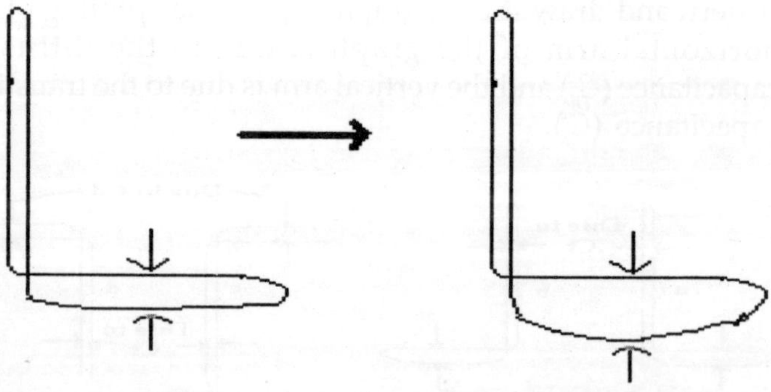

Fig.3 : Doubling of width of the split trace.

$$\frac{C_d}{C_d + C} = \frac{X_1}{X_2} \qquad \text{i.e} \qquad C_d = \frac{C.X_1}{X_2 - X_1} \tag{3.4}$$

For the diode IN 4007 try with C= 0.01, 0.02, 0.03 micro farad, etc., or with any suitable value.

For a forward current the diffusion capacitance ranges from pico farad to a few micro farad.

4
ESTIMATION OF WIDTH OF DEPLETION LAYER

While discussing the Physics and operation of a pn junction diode we come across the terms like depletion layer, barrier potential, donor and acceptor impurity concentration. When a pn junction is formed the majority carriers near the junction crosses the junction creating immobile ions on either sides of the junction. The diffusion of majority carriers across the junction and the formation of depletion layer is depicted in fig.1.

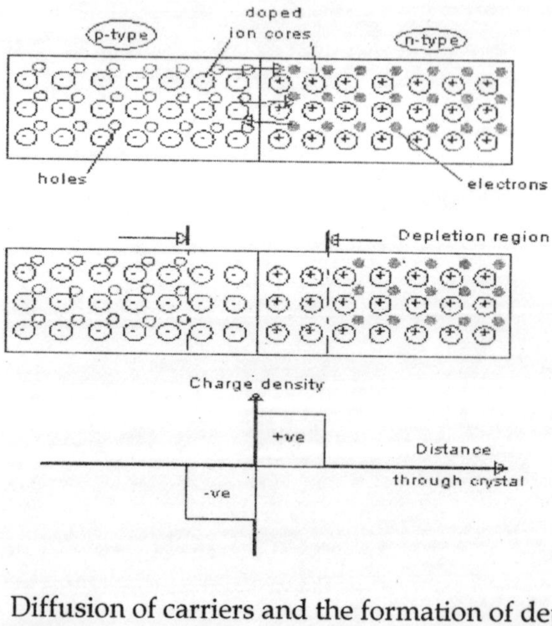

Fig.1 : Diffusion of carriers and the formation of depletion layer.

Thus we get positive immobile ions on the *n* side and negative immobile ions on the *p* side. This migration and neutralisation of majority carriers near the junction continues until the immobile ions on the other side of the junction prevents it. Thus we get a depletion layer at the junction along with the development of a barrier potential gradient along this region. The internal electric field is from *n* to *p* region which is the driving force for the minority carriers. Under steady state the current through the *pn* junction is zero. The equilibrium barrier potential is represented as V_B and the corresponding width of the depletion layer is X_2-X_1.

The dependency of barrier height and width on applied bias, play a crucial role in the function and operation of the resulting semiconductor device. It is convenient to study the properties of one dimensional abrupt homo junctions first and derive expressions for barrier height and width of depletion layer, space charge layer capacitance etc. Since the ideal conditions upon which the expressions have been derived vary from the real situation, the values we compute using such relations can give only values closer to the exact result. This never prevents our understanding the operation of the device.

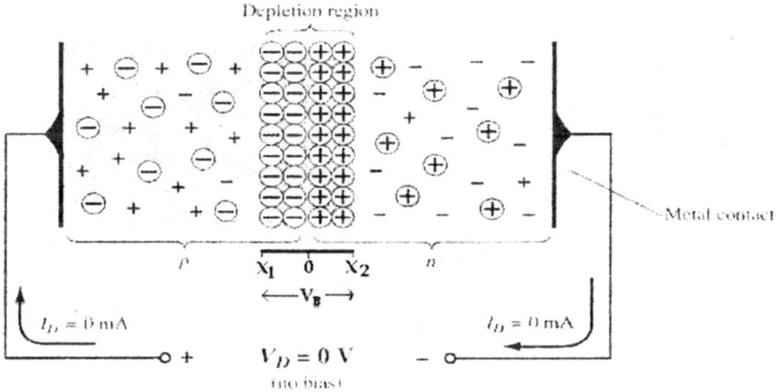

Fig.2 : Barrier potential and depletion layer width under zero bias.

Here we are going to make an approximate estimate of the width of depletion layer from the diffusion capacitance as it dominates in a forward biased diode. One method to determine the width of the depletion layer and the donor impurity concentration is to find it out from the diffusion capacitance. For the determination of diffusion capacitance (C_d) the circuit described in project 3 can be used.

For calculating the width of depletion region (W) we can use the relation

$$W = \frac{\varepsilon_0 \varepsilon_s A}{C_d} \tag{4.1}$$

Where the value of static dielectric constant ε_s is - for (i) silicon – 11.9 (ii) germanium – 16.2 and (iii) gallium arsenide (GaAs) – 13.18. $\varepsilon_0 = 8.854 \times 10^{-12}$ F/m. In arriving at the expression we have made the following assumptions :

(i) depletion layer extends equally on either sides

(ii) the junction is abrupt

(iii) the area of crossection of the junction is equal to the crosssectional area of the diode which is calculated from the radius of the diode.

The radius of the diode is measured with maximum accuracy. For that we can use a traveling microscope with a least count of 0.001 cm. The crosssection of the diode being circular the cross sectional areas

$$A = \pi r^2 \tag{4.2}$$

Thus by finding out the diffusion capacitance by the method described in project 3 and determining the crosssectional area A we can make an estimate of the width of depletion layer.

5
ESTIMATION OF DONOR IMPURITY CONCENTRATION DEBYE LENGTH AND DIFFUSION COEFFICIENT

In order to enhance the conductivity of semiconductor material suitable impurity atoms are added. This process is known as doping. The impurity atoms are of two types donors and acceptors. In a *pn* junction diode we have donor impurity atoms on the *n* side and acceptor impurity atoms on the *p* side. Each donor atom donates an electron for conduction. Under forward bias an electron from each donor impurity atom is let free as conduction electron. Usually the doping concentration ranges from 10^{13} to 10^{18} cm^{-3}. The donor impurity concentration can be calculated from the width of depletion layer.

The width of the space charge region (W) is given by

$$W = \left[\frac{2\varepsilon(V_B - V_a)}{e}\left(\frac{N_a + N_d}{N_a N_d}\right)\right]^{\frac{1}{2}} \tag{5.1}$$

If we assume the acceptor impurity concentration N_a to be approximately equal to donor impurity concentration N_d, we get

$$W = 2\sqrt{\frac{\varepsilon(V_B - V_a)}{eN_d}} \tag{5.2}$$

Where V_B is the builtin barrier potential, V_a - the applied biasing voltage (which is positive ($V_a \leq V_B$) for forward bias and negative for reverse bias), e – the electronic charge and $\varepsilon = \varepsilon_0 \varepsilon_s$; ε_s = 11.8 for silicon. Squaring equation 5.2 and rearranging we get the carrier concentration as

$$N_d = \frac{4\varepsilon(V_B - V_a)}{eW^2} \tag{5.3}$$

Here, the barrier potential V_B can be taken as the knee voltage which can be found from the V-I characteristics of the diode. Thus by determining the barrier potential and width of depletion layer, the donor impurity concentration under zero bias can be determined using the equation 5.4. Under zero bias V_a=0. The value of built in barrier potential V_B can be taken as the knee voltage at which the applied biasing voltage over comes the built in barrier potential.

$$N_d = \frac{4\varepsilon V_B}{eW^2} \tag{5.4}$$

In order to find out the donor impurity concentration, first the diffusion capacitance of the diode is determined by the method described in project 3. Then the width of depletion layer is found out through the method described in project 4. With these data we can make a rough estimate of the donor impurity concentration. In the experiment though we have made several assumptions, it does not prevent us from getting an idea about the carrier concentration. We are bound to make such assumptions for the sake of understanding the Physics involved and experimentation in the labs having limited facilities.

DEBYE LENGTH

This is an important characteristic length for a semiconductor. An important consequence of high doping concentration is that the resultant free carriers screen the Coulomb interaction between the ionized impurity centers and the carriers. Hence the range of the Coulomb interaction decreases with increasing concentration. A screening length is therefore defined to describe this situation which is inversely proportional to free carrier concentration. This screening length is called the Debye length (L_D) and is given by

$$L_D = \sqrt{\frac{\varepsilon kT}{q^2 N}} \tag{5.5}$$

Where q - the charge of the carrier, $\varepsilon = \varepsilon_0 \varepsilon_s$; $\varepsilon_s = 11.8$ for silicon and 16 for germanium, k - the Boltzmann's constant, T - absolute temperature and N the carrier concentration.

At thermal equilibrium the depletion layer width of abruption junctions is about six times the Debye length for germanium, eight times the Debye length for silicon and ten times the Debye length for gallium arsenide. From equation 5.5 we can see that Deby length is a function of doping density. In the case of silicon when the doping concentration changes from 10^{13} cm $^{-3}$ to 10^{19} cm $^{-3}$ the Deby length changes from 10^4 Å to 10^1 Å. (1 angstrom unit Å= 10^{-10}m). For a doping concentration of 10^{16} cm $^{-3}$ the Deby length is about 400 Å. The variation of Debye length as a function of dopping concentration for silicon at 300 K is shown in fig.1.

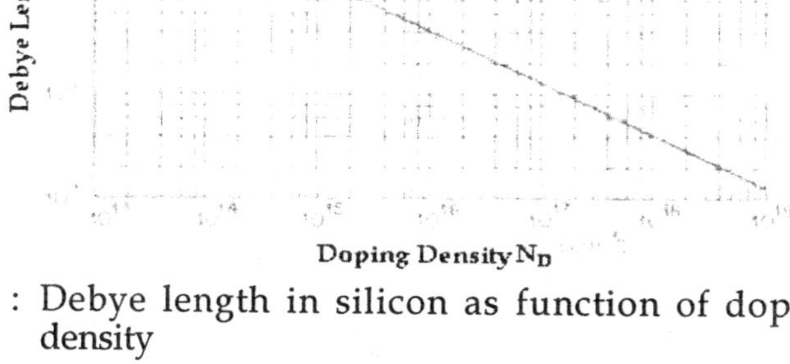

Fig.1 : Debye length in silicon as function of doping density

In the present project work we have already determined the doping concentration using equation 5.4. We can calculate the Debye length using equation 5.5 at a given temperature. One can also make a calculation of the Debye length at different temperatures. What do you observe?

DIFFUSION COEFFICIENT

The flow of carriers from a region of higher number density to a region of lower number density constitutes a current called the diffusion current. The diffusion coefficient (D) depends on temperature (T), impurity concentration and mobility of carriers (μ). Since we have studied the temperature dependence of donor impurity concentration and Debye length, it is worth to calculate the electron and hole diffusion coefficients D_e and D_h.

For the calculation of diffusion coefficient we can use the Einstein's relation

$$\frac{D}{\mu} = \frac{kT}{q} \tag{5.6}$$

Or the electron diffusion coefficient

$$D_e = \frac{kT\mu_e}{q} \tag{5.7}$$

and hole diffusion coefficient

$$D_h = \frac{kT\mu_h}{q} \tag{5.8}$$

Where, q - the charge of the carrier, k - the Boltzmann's constant, T - absolute temperature and μ - mobility of carriers.

Here we can take motilities as given below

Mobility (cm^2/Vs)	Silicon	Germanium
Electron mobility μ_e	1350	3900
Hole mobility μ_h	480	1900

Though this calculation has nothing to do with the experiment it can give a good picture of the variation of diffusion coefficient with temperature.

6
DETERMINATION OF CHARGE OF AN ELECTRON

This is a very simple method for the determination of charge of an electron. While doing the simple experiment V-I characteristics of a diode students do not know the wealth of information hidden in it. They go in search of projects at the end of undergraduate or postgraduate programme without exploring the possibilities of V-I characteristics of a diode, which they have done at the plus two level. This project is also a humble effort to show students how to find out a project.

In this project we are going to use the V-I characteristics of a diode. For this we shall use the experimental set up described in project 1. We have the diode equation for the forward current (I) through a pn junction at temperature T and biasing voltage (V) as

$$I \cong I_0 \left[\exp\left(\frac{eV}{\eta kT} \right) \right] \qquad (6.1)$$

Where η is the ideality factor, k is the Boltzmann's constant and e is the electronic charge. Taking logarithm on both sides of the above equation we get

$$\ln I = \ln I_0 + \left(\frac{e}{\eta kT} \right) V \tag{6.2}$$

If we plot a graph with *ln I* along the *y*–axis and biasing voltage V along the *x*–axis we get a straight line graph

whose slope is $\frac{e}{\eta kT}$.

Hence charge of electron $e = slope.\eta kT$ (6.3)

To carry out the experiment, V-I characteristics of the diode is studied at constant temperature well below the knee voltage as described in project 1. Then plot *ln I-V* graph and determine its slope. Knowing the absolute temperature *T* at which the experiment is carried out one can calculate the electronic charge using equation 6.3 The value of ideality factor can be taken as 2 for silicon and one for germanium for very low biasing voltages.

The result can be further improved if we know the average value of ideality factor under the range of biasing voltage used for the study. This is because for silicon the value of ideality factor ranges from two to one depending upon which current (recombination or diffusion current) dominates. In practical case we cannot isolate one type of current from the other. Hence the value of ideality factor lies between one and two. The variation of ideality factor with forward voltage V_F and current I_F at 300 *K* is shown in fig.1.

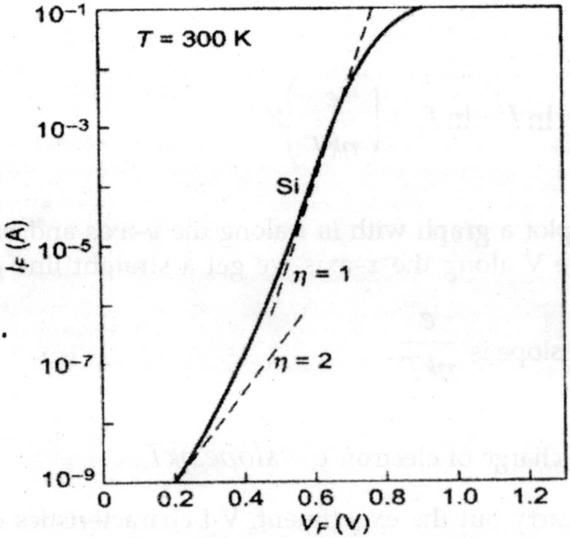

Fig.1 : V-I characteristics of silicon diode at 300K

7
DETERMINATION OF BANDGAP ENERGY

Based on electrical conductivity, materials can be classified as metals, semiconductors and insulators. If we examine the energy level diagram for these materials some characteristic features can be seen.

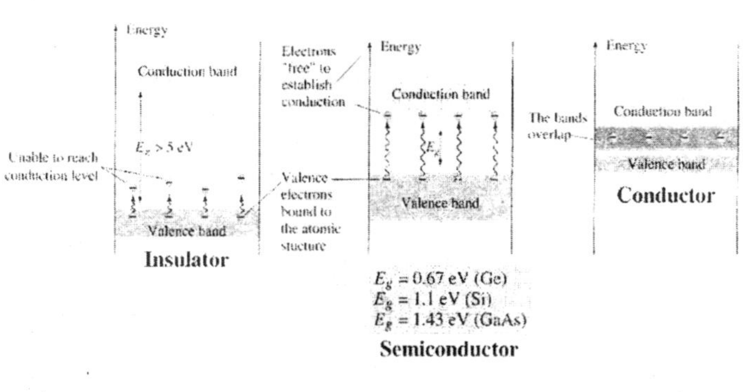

Fig.1 : Energy level diagram for insulator, semiconductor and conductor.

From fig.1 it can be seen that the energy level diagram for insulators and semiconductors contains two bands with a gap in between. It can also be seen that the two bands overlap in the case of metals. The lower energy band which is completely filled is called valance band whereas the upper band which is normally empty is called conduction band. The energy difference between the lower most level in the conduction band and the upper most level in the valance band is called the bandgap energy.

In the case of metals the valance band and the conduction band overlap or there is no energy gap in the energy level diagram. The bandgap energy in insulators is very much greater than that in semiconductors (E_g > 5eV ; see Fig. 1.). Hence is it is very difficult to excite electrons from the valance band to the conduction band in the case of insulators. Semiconductor materials have moderate bandgap energy so that electrons can be easily excited to the conduction band. For this optical, electrical or thermal energy can be used. Thus it is clear that bandgap energy of a material can give information regarding the electrical behaviour of the material. The knowledge of bandgap energy of semiconductor material can help in device applications.

Here I would like to describe one of the simplest method for the determination of bandgap energy (E_g) of a semiconductor material. We know that the forward as well as reverse saturation current depend on temperature. From experimental data it is found that the reverse saturation current (I_o) doubles for every 10 ⁰C rise in temperature.

An empirical relation connecting reverse saturation current, absolute temperature and bandgap energy is given by

$$I_0 = AT^m.\exp\left[\frac{-E_g}{\eta kT}\right] \tag{7.1}$$

Where *A* and *m* are constants, *k* the Boltzmann's constant and h the ideality factor. For germanium η=1 , *m*=2 and for silicon η=2 and *m*=1.5.

For a forward biased diode the current is given by

$$I = I_0\left[\exp\left(\frac{eV}{\eta kT}\right) - 1\right] \tag{7.2}$$

where I_0 is the reverse saturation current. Neglecting one in the above equation we get the reverse saturation current as

$$I_0 = I\left[\exp\left(\frac{-eV}{\eta kT}\right)\right] \tag{7.3}$$

For a germanium diode the reverse saturation current is high compared to silicon diode. Hence it can be measured directly at different temperature by studying the reverse bias characteristics. But in the case of silicon diode the reverse current is very low and is difficult to measure using and ordinary multimeter. For getting the reverse saturation current in a silicon diode high reverse bias is required. This results in the flow of considerable surface leakage current making it difficult to measure the reverse saturation current. Hence in the case of silicon diode the forward current is measured at different temperature and from this data the reverse saturation current is calculated using the relation 7.3.

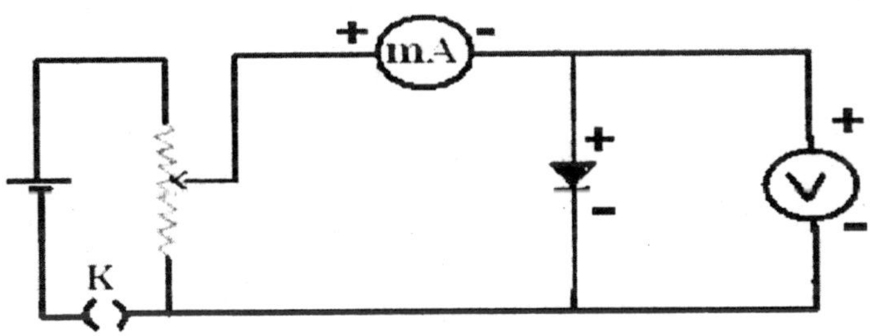

Fig. 1 : Circuit to study the forward characteristics of a silicon diode.

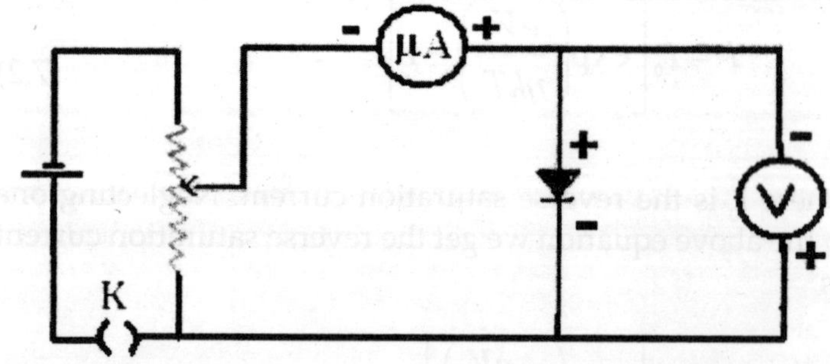

Fig. 2 : Circuit to study the reverse characteristics of a germanium diode.

Using circuit shown in fig.1 we can determine the forward current (I) for various biasing voltages for a silicon diode. For germanium diode, using the circuit shown in fig. 2 we can find the reverse saturation current (I_0) for various biasing voltages. In order to study the effect of temperature on diode current (forward or reverse) the diode is heated indirectly by keeping it in a pure water bath or oil bath (use transformer oil).

Let the reverse saturation current at temperature T_1 and T_2 be I_{01} and I_{02} respectively. By equation 7.1

$$I_{01} = AT_1^{\,m}.\exp\left[\frac{-E_g}{\eta kT_1}\right] \qquad (7.4)$$

$$I_{02} = AT_2^{\,m}.\exp\left[\frac{-E_g}{\eta kT_2}\right] \qquad (7.5)$$

Dividing equation 7.4 by equation 7.5

$$\frac{I_{01}}{I_{02}} = \frac{T_1^m}{T_2^m} \exp\left[\frac{E_g}{\eta k}\left(\frac{1}{T_2} - \frac{1}{T_1}\right)\right]$$

Taking logarithm on both sides

$$\ln\left[\frac{I_{01}}{I_{02}} \frac{T_2^m}{T_1^m}\right] = \left[\frac{E_g}{\eta k}\left(\frac{1}{T_2} - \frac{1}{T_1}\right)\right]$$

Or

$$\left[\ln I_{01} - m \ln T_1\right] - \left[\ln I_{02} - m \ln T_2\right] = \frac{E_g}{\eta k}\left[\frac{1}{T_2} - \frac{1}{T_1}\right]$$

Equation 7.6 shows that if we plot a graph with (ln I_0 - m ln$T$) along the y-axis and (1/T) along the x-axis we get a straight line graph with slope (E_g/hk).

i.e. $E_g = slope.\eta k$ joules (7.7)

For silicon I_0 is calculated using equation 7.3 and (ln I_0 -1.5ln$T$) is found at different temperature T. In the case of germanium since we can measure I_0 directly for plotting graph we need find only (ln I_0 -2ln$T$) only. (Remember m=1.5 for silicon and 2 for germanium). The observations can be tabulated as follows.

Table 1 : For germanium (Reverse characteristics)

Temperature T (K)	Reverse biasing voltage (V)	Reverse current (μA)	1/T	(ln I_0 -2lnT)

Table 2 : For silicon (Forward characteristics)

Temperature T (K)	Forward biasing voltage (V)	Forward current I (mA)	$\ln I_0$ *	1/T	$(\ln I_0 - 1.51nT)$

where * $\ln I_0 = \ln I - \left(\dfrac{eV}{2kT}\right)$

This can also be determined graphically. Plot a graph with $\ln I$ along the y-axis and V along the x-axis for each temperature. The y-intercept of the graph gives the value of $\ln I_0$.

From these observations and graph the bandgap energy can be calculated using equation 7.7 . To get the bandgap energy in electron volt (eV) you have to divide equation 7.7 by the electronic charge e. Instead of taking the value of ideality factor for silicon as two, if the actual value (which can be determined by the method described in project 2) is substituted in equation 7.7 the unnecessary adjustments in the slope of the graph (that students usually do) can be avoided.

8
DETERMINATION OF DONOR IMPURITY CONCENTRATION AND WIDTH OF DEPLETION LAYER FROM FORWARD CHARACTERISTICS

In order to enhance the conductivity of a semiconductor certain impurity atoms are added. This process is called doping. If the impurity atom donates an electron to the crystal it is called donor impurity and if the impurity atom creates a vacancy in the covalent bond it is called acceptor impurity. When doped semiconductor is used to form a *pn* junction, each impurity atom contributes an electron / hole to the crystal for conduction. The number of free charge carriers in the crystal increases with the increase of the forward biasing voltage given to the diode. We experience this while studying the V-I characteristics as an increase in diode current.

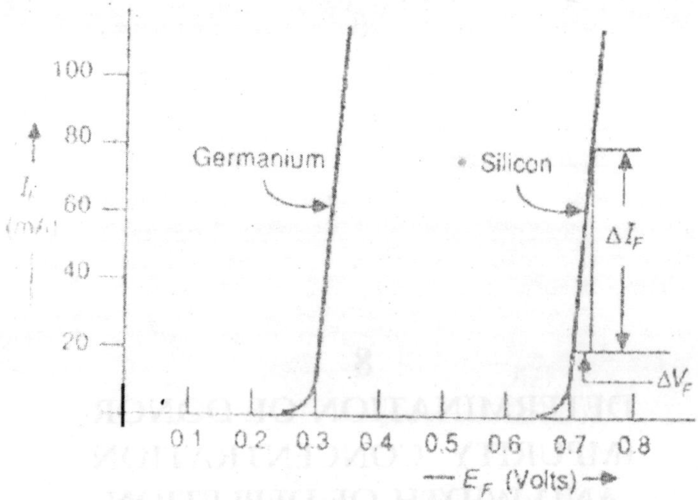

Fig. 1 : Forward characteristics of germanium and silicon diode.

From the V-I characteristics shown in fig. 1 one can see that as the forward biasing voltage increases the forward current also increases. As mentioned earlier, the increase of current is due to the release of more and more carriers for conduction mainly from impurity atom. There is nothing wrong in assuming that the electron density (n) is approximately equal to donor impurity concentration (N_d).

While deriving the diode equation we have used the relation for the current (I) as

$$I = neAv_d = N_d eAv_d \tag{8.1}$$

Where A is the area of crosssection of the semiconductor diode and v_d the drift velocity of the carrier (electron) of charge e. We know that the drift velocity of electron is related to the mobility (m) of the carrier and the force F acting on it by the relation,

$$v_d = \mu F \tag{8.2}$$

$\therefore$ Equation 8.1 becomes $I = N_d e A \mu F$ (8.3)

The force (*F*) acting on a charge (q) = *F*= *qE*

Or

Force acting on unit charge = *F/q* = *E*, the electric field acting on the charge. Therefore equation 8.3 becomes

$$I = N_d e A \ \mu E = N_d e A \ \mu V/d \tag{8.4}$$

Since *E*= *V/d* . Where *V* is voltage applied across the diode of length d. From equation 8.4 the donor impurity concentration can be written as

$$N_d \cong n = \frac{I.d}{eAV\mu} = \frac{d}{eA\mu R} \tag{8.5}$$

Where *R* = *V/I* is the diode resistance and *A* = πr^2 is the area of crosssection of the diode of radius r and length *d*. The mobility of electron in silicon and germanium crystal is given in Appendix A.

The radius (*r*) and length (*d*) of the diode can be accurately measured using a traveling microscope (if not available, use a screw gauge). Thus, we can find the area of crosssection (*A*) of the diode. For studying the *V-I* characteristics of the diode we can use the forward biasing circuit given in project 1. The forward current (*I*) is measured for different biasing voltages (*V*). The value of *R*=*V/I* is calculated in each case. Using equation 8.5 we can calculate donor impurity concentration for various biasing voltage.

Now a doubt may arise. How can the donor impurity concentration vary with biasing voltage? For a given diode the donor impurity concentration is constant. The value given by equation 8.5 is actually the carrier concentration. We have made an assumption that free electron concentration is approximately equal to the donor impurity concentration. i.e. $n \approx N_d$. For low biasing voltage due to barrier potential the diode current is low. Hence the value of carrier concentration, $n \approx N_d$, we get also is low. Beyond the cutin voltage or knee voltage the semiconductor diode behaves almost like a conductor. Hence we get a near linear portion beyond the cutin voltage. In this region, since $\Delta V/\Delta I$ can be considered nearly as a constant, the value of $n \approx N_d$ also becomes a constant. The *V-I* characteristics of a diode is shown in fig. 2.

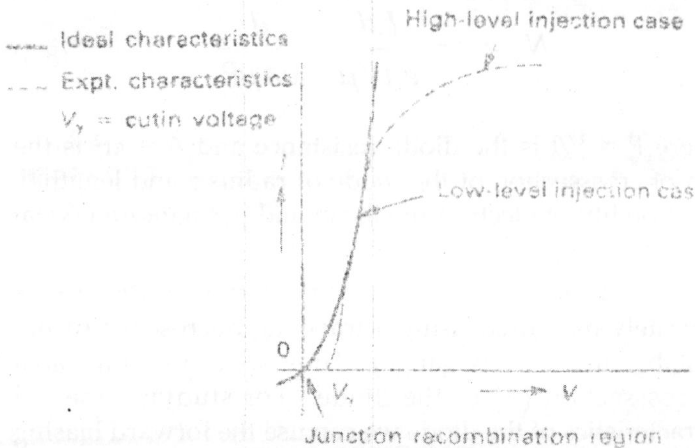

Fig. 2 : The V-I characteristics of a diode.

As we increase the biasing voltage we move into the high level injection region of the *V-I* characteristics. Here, in addition to the electrons from the donor atoms bond electrons of the crystal also start contributing to the diode current. Hence the value of diode resistance in this region cannot give the value for donor impurity concentration.

Thus it will be good to find the value of diode resistance at the knee voltage or just above the knee voltage. The value of carrier concentration (n) given by equation 8.5 can be taken as the approximate value for the donor impurity concentration (N_d).

Then a graph is drawn with the biasing voltage along the x-axis and N_d value along the y-axis. We can see how the value of carrier concentration varies with biasing voltage. It will be difficult to see a saturation region in the graph as the contribution of bond electrons also come into play with the increase of biasing voltage. Hence we can take the value of N_d at the knee voltage, as the donor impurity concentration. Or, simply knowing the cutin voltage and the then forward current, we can compute the value of donor impurity concentration using equation 8.5. Usually the doping concentration ranges from 10^{13} cm^{-3} to 10^{18} cm^{3}.

WIDTH OF DEPLETION LAYER

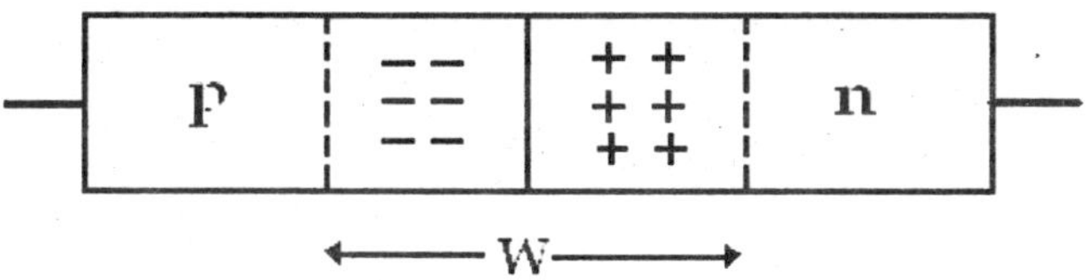

Fig. 3 : pn junction with a depletion layer of width *W*.

When a pn junction is formed, due to the diffusion of charge carriers across the junction a region deprived of majority carriers is formed on either sides of the junction. Thus negative ions are formed on the p-side and positive ions are formed on the n-side (see fig. 3). The diffusion process continues until the negative ions on the p-side repel the additional electrons trying to cross from the n–side and the positive ions on the n–side repel the additional holes trying to cross from the p–side. Thus the initial diffusion of charge carriers creates a barrier potential at the junction. The typical barrier potential (V_B) is 0.3 V for germanium and 0.7 V for silicon. When no biasing voltage is applied the value of barrier potential (V_B) depends on temperature and doping. The region of immobile ions at the junction is known as depletion layer or space charge region or the transition region. The width of this region is called barrier width or width of depletion layer. The width is of the order of microns. For light doping the width of depletion layer is very large and it decreases with doping concentration.

The width of the depletion layer can be estimated form the *V-I* characteristics of the diode. The width of the space charge region (*W*) is given by

(8.6)

If we assume the acceptor impurity concentration $N_a \approx N_d$ donor impurity concentration, we get

$$W = 2\sqrt{\frac{\varepsilon(V_B - V_a)}{eN_d}} \qquad (8.7)$$

Where V_B is the builtin barrier potential, V_a - the applied biasing voltage (which is positive ($V_a \leq V_B$) for forward bias and negative for reverse bias), e – the electronic charge and $\varepsilon=\varepsilon_0\varepsilon_s$; ($\varepsilon_s$ = 11.8 for silicon). When $V_a=V_B$, $W=0$. This shows beyond the knee voltage the width of space charge region reduces to zero and the current shoots up.

We can find the barrier potential from the *V-I* characteristics. When a diode is forward biased the current first increases gradually and then sharply. This is because, the applied voltage is used to overcome the internal potential barrier. The forward voltage at which the current starts to increase rapidly is known as the cutin voltage or knee voltage. This can be taken as the value for barrier potential (V_B). Knowing the current through the junction at the forward voltage V_B the doping concentration (N_d) is determined using the equation 8.5. With this value of V_B and N_d the width of depletion layer can be determined using the relation 8.8.

$$W = 2\sqrt{\frac{\varepsilon V_B}{eN_d}} \qquad (8.8)$$

DEBYE LENGTH

We have already discussed Debye length in project 5. From the value of doping concentration calculated using equation 8.5, we can calculate the Debye length using the relation

$$L_D = \sqrt{\frac{\varepsilon kT}{q^2 N}} \qquad (8.9)$$

Where q - the charge of the carrier, $\varepsilon = \varepsilon_0\varepsilon_s$; $\varepsilon_s = 11.8$ for silicon and 16 for germanium, k - the Boltzmann's constant, T - absolute temperature and N the carrier concentration.

9
DETERMINATION OF BANDGAP ENERGY - USING WHEATSTONE'S BRIDGE

We know that the energy spectrum of a semiconductor consists of valance band and conduction band. At absolute zero temperature the valance band is completely filled and conduction band is completely empty. Thus due to the non availability of free electrons in the conduction band the semiconductor behaves like an insulator.

When the temperature is raised some of the covalent bonds in the semiconductor breaks and releases electrons to become free in the conduction band. The availability of free electrons in the conduction band improves the conductivity of the crystal.This shows that semiconductors have negative temperature coefficient of resistance.

In the case of a *pn* junction diode the conductivity can be increased either by raising the temperature or by increasing the forward bias. Thus the resistivity (ρ) of the diode decreases with the raise of temperature in manner represented by the equation given bellow.

$$\rho = \rho_0 e^{B/T} \tag{9.1}$$

Where ρ_0 is the resistivity at temperature $T=0$ K and B a constant. If we assume that the dimension of the diode does not change on heating, we can replace ρ by its resistance R. Therefore equation 9.1 becomes

$$R = R_0 e^{B/T} \tag{9.2}$$

$$\ln R = \ln R_0 + \frac{B}{T} \tag{9.3}$$

If we plot ln *R* along the y axis and (*1/T*) along the x-axis, we get a straight line graph with a positive slope *B*. For a semiconductor the constant *B* is given by

$$slope = B = \frac{E_g}{\eta k} \tag{9.4}$$

Where η is the ideality factor, E_g the bandgap energy and *k* the Boltzmann's constant. From equation 9.4 the bandgap energy can be calculated as

$$E_g = \eta . k . slope \qquad \text{joules} \tag{9.5}$$

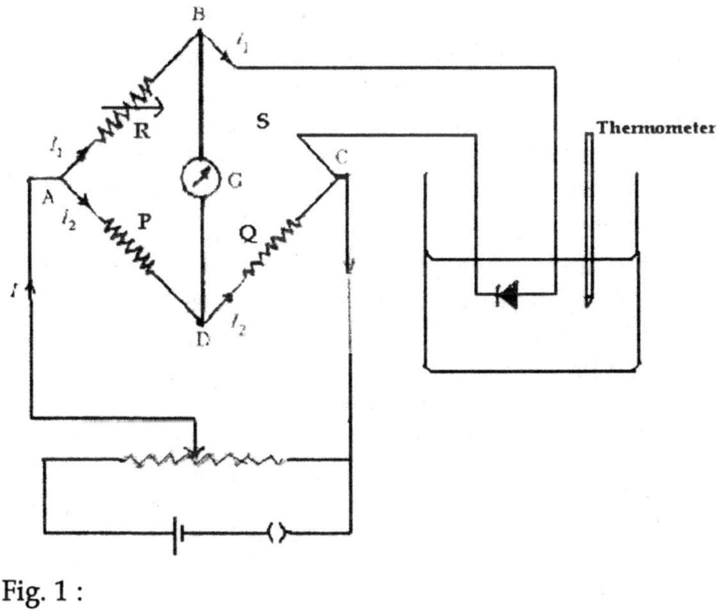

Fig. 1 :

To determine the resistance of the diode the principle of Wheatstone's bridge can be used. A Wheatstone's bridge is constructed as shown in fig. 1 with the diode in one arm (say *S*). The resistance *R* and *S* are set equal (say 100Ω, 200 Ω, 1000 Ω or 2000Ω). Since $P = Q$, the value of *R* should be adjusted to the diode resistance for getting the balancing condition. When the bridge is balanced the galvnometer shows null deflection.

Then,

$$\frac{P}{Q} = \frac{R}{S} = 1$$

Or $R = S$ where S represents the diode resistance. Under balanced condition the value of *R* will give the diode resistance.

We know that the diode resistance varies from a high value under low forward bias to a low value beyond the knee voltage. Hence it is possible to set the resistance of the diode to any value by suitably varying the biasing voltage. If we fix the value of *R*, say to 100 Ω, 200Ω, 1000Ω or 2000 Ω, we have to adjust the rheostat for getting the bridge balanced. i.e. In the present work, I would like to fix the value of R and adjust the rheostat for getting the balancing condition. Thus the value of diode resistance at room temperature is known. Use a fractional resistance box in the place of *R*.

If the diode is heated by keeping it in an oil bath or pure water bath, the diode resistance decreases and the bridge becomes unbalanced. Now by suitably changing the value of *R* the bridge can be balanced. This value of *R* gives the value of diode resistance at that temperature. Thus we can measure the diode resistance ($R=S$) at various temperatures. Then a graph is plotted with ln R along the yaxis and ($1/T$) along the xaxis. We get a straight line graph as shown in fig. 2. From the slope of the graph the bandgap energy can be determined using equation 9.5.

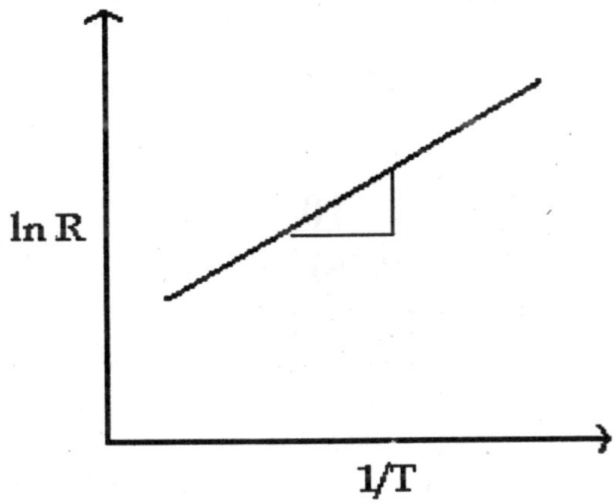

Fig. 2 : ln R - $1/T$ graph

If we know the diode resistance for various biasing voltages, it is possible to carry out the experiment at any desired region of the *V-I* characteristics. Since the biasing voltage is kept constant the variations in diode resistance is due to thermally generated current. This is one of the simplest methods for the determination of bandgap energy of a semiconductor taken in the form of a diode.

10
DYNAMIC RESISTANCE AND IDEALITY FACTOR

I hope you are enjoying the beauty of the simple diode characteristics. The simple projects using a pn junction diode described so far are capable of giving you a better understanding of the Physics of pn junction. Here is another simple method to study the contribution of recombination current and diffusion current to the total diode current. In other words, the present project will help you to check which current dominates at particular temperature and biasing voltage. The value of ideality factor is capable of telling you that. Ideal diffusion recombination and total current in a forward - biased pn junction is shown in figure 1.

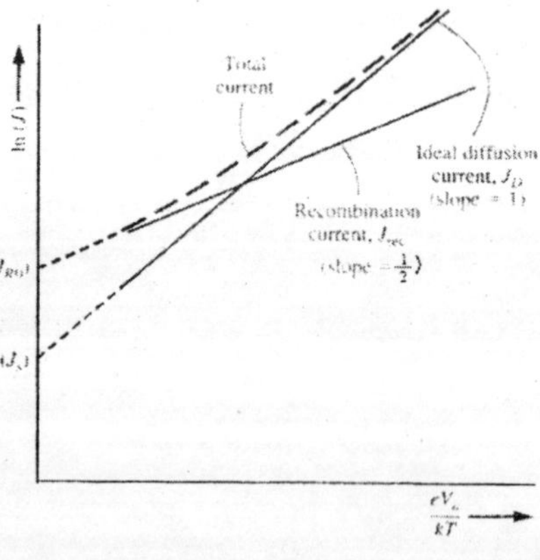

Fig. 1 : Ideal diffusion, recombination and total current in a forward+biased *pn* junction.

In this project we are going to determine the dynamic resistance of the diode and ideality factor from the forward characteristics. For studying the forward characteristics let us use the circuit given in fig. 2.

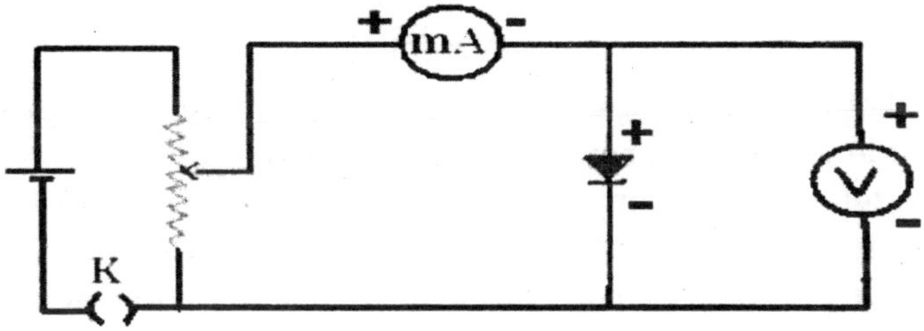

Fig. 2 : Circuit to study the forward characteristics of a diode

The forward current is noted by changing the biasing voltage. We get the *V-I* characteristics as given below (fig. 3).

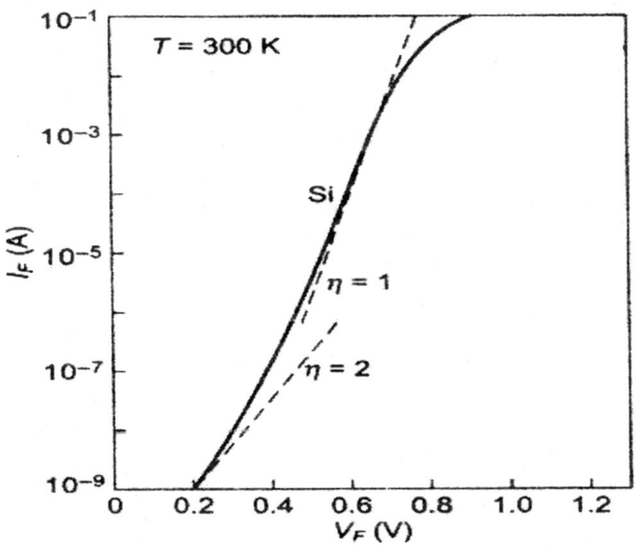

Fig. 3 : V-I characteristics of silicon diode at 300K

The important diode parameters are :

1. forward voltage drop
2. reverse saturation current
3. reverse break down voltage
4. dynamic resistance
5. maximum forward current

The value of these quantities is normally listed in the diode data sheet provided by the manufacturer. These can be determined from the forward or reverse characteristics. Students are also advised to make an effort to study the above parameters. I am sure that each study will teach you something. Only one thing you have to keep in mind is that never make a foolish attempt of classifying a topic as plus two (intermediate) level or undergraduate level or post graduate level. Remember, there is one and only Physics for all classes and we explore it from the level of our understanding. The greater our level of understanding the greater is the information we deduce form small things.

In the present project work we are going to find out the knee voltage and dynamic resistance from the forward characteristics. Then from these data we will be calculating the ideality factor. The dynamic resistance (r_d) is also known as the incremental resistance or ac resistance. It is defined as the reciprocal of the slope of the *V-I* characteristics beyond the knee.

$$r_d = \frac{dV}{dI} \tag{10.1}$$

We have the diode equation

$$I \cong I_0 \left[\exp\left(\frac{eV}{\eta kT} \right) \right] \tag{10.2}$$

Taking logarithm on both sides,

$$\ln I = \ln I_0 + \left(\frac{eV}{\eta k}\right)\frac{1}{T} \tag{10.3}$$

Differentiating equation 10.3

$$\frac{1}{I}dI = \frac{e}{\eta kT}dV$$

Or

$$\frac{dV}{dI} = \frac{\eta kT}{Ie}$$

Substituting the values of e and k we get

$$r_d = \frac{\eta T}{11600I} \tag{10.4}$$

Since $(e/k) = 11594 \approx 11600$

Or $$\eta = \frac{11600 I r_d}{T} \tag{10.5}$$

At 300K $$\eta = 38.6 r_d I \tag{10.6}$$

Although r_d varies with current, in a small signal model, it is reasonable to use the parameter r_d as a constant beyond the knee. In order to find the ideality factor, the forward characteristics of the diode is studied as described in project 1.

From the forward characteristics one can see that there exist a cutin /offset /break point/threshold/knee voltage below which the current is very low and beyond which it rises rapidly almost like a conductor. The knee voltage is approximately 0.3 V for germanium and 0.6 - 0.7 V for silicon. The cutin or knee voltage can be found from the V-I characteristics as shown in fig. 4.

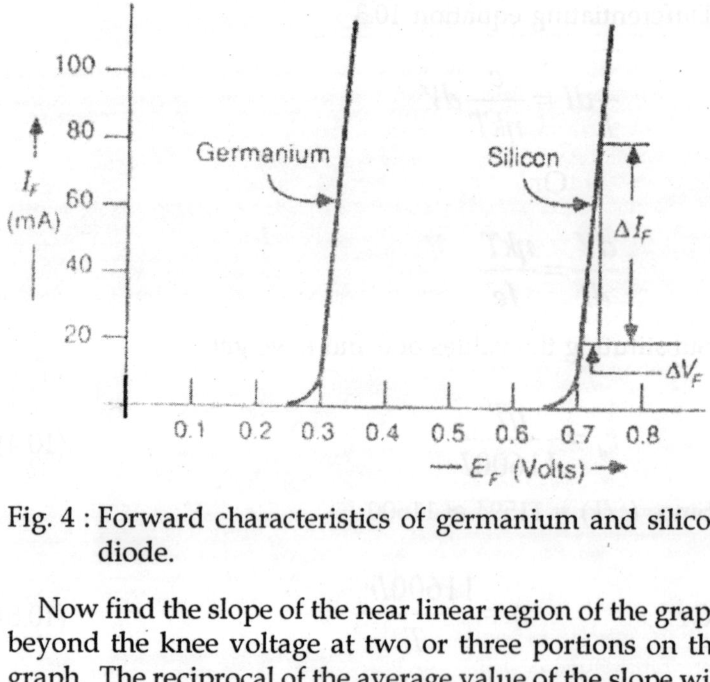

Fig. 4 : Forward characteristics of germanium and silicon diode.

Now find the slope of the near linear region of the graph beyond the knee voltage at two or three portions on the graph. The reciprocal of the average value of the slope will give the dynamic resistance of the diode. Knowing the dynamic resistance the ideality factor can be found using the relation 10.5. As shown in fig. 1 and 3 in this region of the *V-I* characteristics the diffusion current dominates. Hence we will be getting the value of ideality factor between one and two.

Do the experiment with different diodes like BA 159, BY399, IN 5402, IN 5408, IN 4001 - 4007, etc. Do you find any change? Also extend the project work to study whether there is any effect of temperature on ideality factor. See how the simple *V-I* characteristics has turned out a beautiful project work.

11
DETERMINATION OF BANDGAP ENERGY USING INSTRUMENTATION AMPLIFIER

The bandgap energy of a solid determines whether it is an insulator, conductor or a semiconductor. The energy difference between the valence band and the conduction band is called the bandgap energy (E_g). The important semiconductors silicon and germanium have bandgap energy 1.2 eV and 0.785 eV respectively at 0 K.

So far we have discussed two methods for the determination of bandgap energy. The project 9 has used a Wheatstone's bridge for the determination of bandgap energy. There we determined the diode resistance directly by balancing the bridge by adjusting the variable resistance arm. Throughout the work the biasing voltage is kept constant. Now we are going to discuss another method of using an instrumentation amplifier for the determination of bandgap energy.

We know that when the temperature of the diode changes the diode current also changes. If we connect the diode as part of a balanced bridge, as given in fig. 1, we can see that the balancing condition is lost on heating the diode. This is due to the change in resistance of the diode with temperature. Hence the potential difference across the points *AB* changes from the initial value zero. This potential difference is fed to an instrumentation amplifier whose output will be proportional to the change in potential difference which in turn is proportional to the change in resistance of the diode.

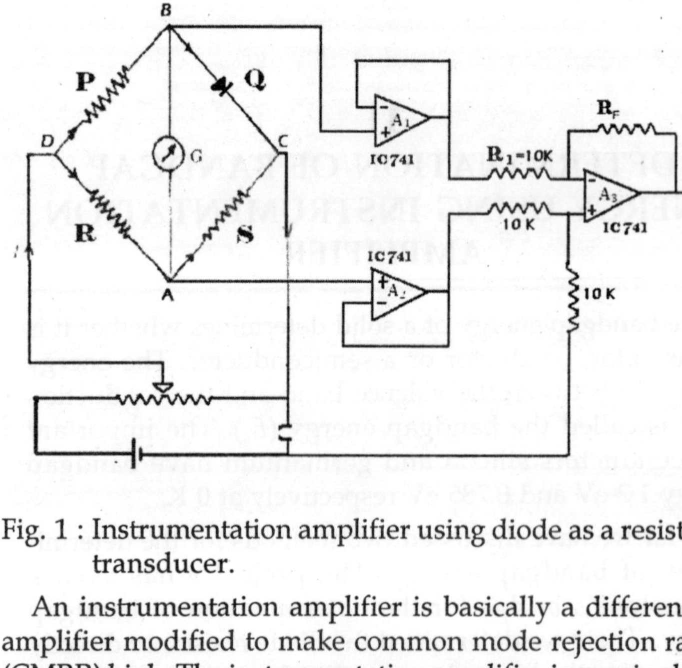

Fig. 1 : Instrumentation amplifier using diode as a resistive transducer.

An instrumentation amplifier is basically a differential amplifier modified to make common mode rejection ratio (CMRR) high. The instrumentation amplifier is required for precise low level signal amplification. They are used where low noise, high input resistance, accurate closed loop gain and high common mode rejection ratio (CMRR) are required. The high input impedance of the instrumentation amplifier increases the differential mode gain. This is the reason why instrumentation amplifiers are widely used in data acquisition systems, instrumentation systems to measure temperature, humidity, light intensity, etc. Most of the instrumentation amplifier uses a transducer in a bridge circuit. It facilitates the amplification of potential difference appearing at *AB* (see Fig.1) due to the imbalance of the bridge. The imbalance of the bridge arises due to the change in the physical quantity.

The figure shows a simplified differential instrumentation amplifier using a resistive transducer bridge. Here the resistive transducer is a silicon diode whose bandgap energy is to be determined. Generally, the resistors P, R and S are selected so that they are equal in value to the transducer resistance Q at some reference condition. The reference condition is the specific value of the physical quantity (the resistance of the diode at a given temperature, say room temperature) under measurement at which the bridge is balanced., i.e. When the bridge is balanced we can write $Q=P=S=R$.

When temperature changes by ΔT, the resistance of the transducer (here, diode) also changes by ΔR. Then the potential difference across the output terminals of the bridge is given by

$$V_{ab} = \frac{-\Delta R.V_{DC}}{2(2R + \Delta R)} \qquad (11.1)$$

This voltage is applied to the instrumentation amplifier. The voltage followers preceding the basic differential amplifier help to eliminate the loading effect of the circuit. The gain of the differential amplifier is $(-R_F/R_1)$. Therefore the output of the instrumentation amplifier is

$$V_0 = V_{ab}\left(\frac{-R_F}{R_1}\right) = \frac{\Delta R V_{DC} R_F}{(4R + 2\Delta R)R_1}$$

For small ΔR, the above equation can be written as

$$V_0 = \frac{R_F \Delta R V_{DC}}{4R.R_1} \qquad (11.2)$$

This shows that the output voltage is proportional to the change in resistance ΔR of the diode.

The resistivity (ρ) of the diode decreases with the raise of temperature in a manner given by the equation given bellow.

$$\rho = \rho_0 e^{B/T} \qquad (11.3)$$

Where ρ_0 is the resistivity at temperature *T*=0 *K* and *B* a constant. If we assume that the dimension of the diode does not change on heating we can replace ρ by its resistance *R*. Therefore equation 9.1 becomes

$$R = R_0 e^{B/T} \qquad (11.4)$$

$$\ln R = \ln R_0 + \frac{B}{T} \qquad (11.5)$$

When temperature changes the resistance of the diode also changes. From equation 11.2 it can be seen that output voltage of the instrumentation amplifier is proportional to the change in resistance ΔR of the diode. Hence the above equation 11.4 and 11.5 can be modified as

$$V = V_0 e^{B/T}$$

$$\ln V = \ln V_0 + \frac{B}{T}$$

If we plot ln V_0 along the y axis and (1/*T*) along the x-axis we get a straight line graph with a positive slope *B*. The slope *B* is given by

$$slope = B = \frac{E_g}{\eta k} \qquad (11.6)$$

Where η is the ideality factor, E_g the bandgap energy and k the Blotzmann's constant. From equation 9.4

the badgap energy can be calculated as

$$E_g = \eta.k.slope \text{ joules}$$

$$E_g = \frac{\eta.k.slope}{e} = \frac{\eta.slope}{11600} \text{eV} \qquad (11.7)$$

If the inputs of differential amplifier are interchanged we get opposite sign for the output voltage. There will not be any change in the slope.

Since we have to use the value of ideality factor η, it is better to determine its value by the method described in project 2. To determine the bandgap energy of the diode the diode is connected in one of the arms of the bridge as shown in fig. 1. The value of the resistors *P*, *R*, and *S* are set equal (say 500 ohm, 1 k ohm etc). The voltage applied to the bridge is varied until the resistance (*Q*) of the diode becomes equal to the resistances in the other arms of the bridge. The balancing condition can be checked by null deflection method using a galvanometer connected between *A* and *B*. Then remove the galvanometer and connect the junctions to the inputs of the op-amp A_1 and A_2 of the instrumentation amplifier. Keeping the diode in an oil bath or pure water bath vary the temperature of the bath and note the output voltage of the instrumentation amplifier. Then plot ln V_0 - $(1/T)$ graph. We get a straight line graph with a slope *B*. Determining the slope of graph the bandgap energy can be calculated using the relation 11.7.

This work is an attempt to determine the bandgap energy of semiconductor using instrumentation amplifier. The method can be employed for the detection of small changes in the diode current as a result of temperature variation. The circuit also uses the privilege of balanced bridge and instrumentation amplifier.

12
ANOTHER METHOD FOR THE DETERMINATION OF BANDGAP ENERGY

So far we have discussed two methods for the determination of bandgap energy in project 7 and 11. Here I would like to discuss one more method of manipulating the V-I characteristics for the determination of bandgap energy. In this method we will be using two different formulae for germanium and silicon.

For Germanium

We have the diode equation

$$I \cong I_0\left[\exp\left(\frac{eV}{\eta kT}\right)\right] \tag{12.1}$$

Where I is the current through the forward biased pn junction at temperature T (in Kelvin scale) and biasing voltage V. I_0, e, k and η denote the reverse saturation current, electronic charge, Boltzmann's constant and ideality factor respectively.

An empirical relation connecting reverse saturation current, absolute temperature and bandgap energy is given by

$$I_0 = AT^m.\exp\left[\frac{-E_g}{\eta kT}\right] = AT^m \exp\left[\frac{-E_g}{\eta V_T}\right] \tag{12.2}$$

Where A and m are constants, k the Boltzmann's constant, η the ideality factor, V_T is the volt equivalent of temperature defined as kT/e (or T/11600) and V_{GO} is the voltage which is numerically equal to the bandgap energy E_g in electron volt at 0K. For germanium η=1 , m=2 and for silicon η=2 and m=1.5.

Taking logarithm on equation 12.2

$$\ln I_0 = C - \frac{E_g}{kT} \tag{12.3}$$

Where $C=\ln(AT^2)$. Equation 12.3 shows that if we plot a graph with $\ln I_0$ along the y axis and $(1/T)$ along the xaxis we get a straight line graph with slope (E_g/k).

i.e. $E_g = slope.k$ J

or $E_g = slope.k/e$ eV (12.4)

To carry out the experiment the circuit given as fig. 2 in the project 7 can be used. Measure the reverse saturation current using a micro ammeter or digital multimeter for different biasing voltages. Repeat the experiment for different temperatures. Then from the $\ln I_0$ - $(1/T)$ graph the bandgap energy can be calculated using equation 12.4.

For silicon

We can determine bandgap energy using equation 12.1 and 12.2. Taking logarithm on equation 12.2 and differentiating we get,

$$\frac{1}{I_o}\frac{dI_0}{dT} = \frac{d(\ln I_0)}{dT} = \frac{m}{T} - \frac{E_g}{\eta}\frac{d}{dT}\left(\frac{1}{V_T}\right)$$

where V_{GO} is written as E_g in eV.

$$\frac{1}{I_o}\frac{d(\ln I_0)}{dT} = \frac{m}{T} + \frac{E_g}{\eta T V_T} \tag{12.5}$$

Taking logarithm on equation 12.1

$$\ln I = \ln I_0 + \left(\frac{eV}{\eta k}\right)\frac{1}{T} \tag{12.6}$$

Differentiating equation 12.6 for constant I and substituting equation 12.5, we get

$$\frac{dV}{dT} = \frac{V}{T} - \eta V_T\left(\frac{1}{I_0}\frac{dI_0}{dT}\right) = \frac{V - (E_g + m\eta V_T)}{T} \tag{12.7}$$

Or

$$E_g = V - m\eta V_T - T\frac{dV}{dT} \tag{12.8}$$

The diode current is varied using the circuit shown in fig. 1. For a given diode current measure the voltage across the diode at different temperature. Repeat the experiment for different diode current. Then plot a graph with voltage along the yaxis and temperature along the x-axis. From the slope of the graph dV/dT can be determined. From this graph for a given temperature, the voltage across the diode can be found and the bandgap energy can be calculated using equation 12.8. Remember the accuracy can be improved if we know the value of ideality factor in the operating region.

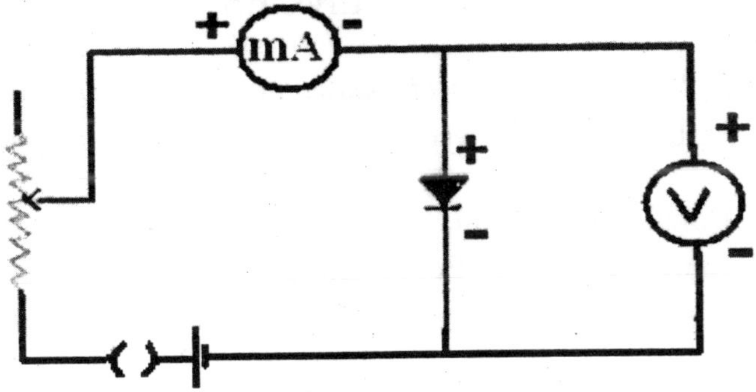

Fig. 1 : Circuit to study forward characteristics of silicon diode.

REFERENCES

1. S. M. Sze, *Physics of Semiconductor Devices*, Wiley Eastern Limited, Second edition (1993).

2. Donald A Neaman, *Semiconductor Physics and Devices*, Tata McGraw Hill Publishing Company Limited, Third edition (2002).

3. Dilip K Roy, *Physics of Semiconductor Devices*, University Press, Second edition (2004).

4. S. S. Islam, *Semiconductor Physics and Devices*, Oxford University Press (2008).

5. J. Millman and C. C. Halkias, *Integrated Electronics: Analog and Digital Circuits and Systems*, Tata McGraw Hill Publishing Company Limited (1991).

6. David A Bell, *Electronic Devices and Circuits*, Prentice Hall of India Private Limited, Third edition (2000).

7. S. Sankararaman, LE , *J. Lab Experiments*, 4(3) (2004) 174.

8. S. Sankararaman, LE , *J. Lab Experiments*, 6(1) (2006) 32.

APPENDIX

APPENDIX A

PROPERTIES OF SILICON, GERMANIUM AND GALLIUM ARSENIDE

Property	Silicon	Germanium	Gallium Arsenide
Atoms ($\times 10^{22}/cm^3$)	5	4.42	4.42
Atomic weight	28.09	72.6	144.63
Crystal structure	Diamond	Diamond	Zincblende
Density (g/cm^3)	2.33	5.33	5.32
Dielectric constant	11.7	16.0	13.1
Bandgap energy at 300K	1.12	0.66	1.42
Intrinsic carrier concentration (cm^{-3})	1.45×10^{10}	2.4×10^{13}	1.79×10^{6}
Intrinsic resistivity (Ω - cm)	2.3×10^{5}	47	3.3×10^{8}
Mobility (cm^2/V-s)			
Electron (μ_e)	≤ 1400	≤ 3900	≤ 8500
Hole (μ_h)	≤ 480	≤ 1900	≤ 400
Specific heat (J/g-^0C)	0.7	0.31	0.33
Thermal conductivity at 300K(W/cm-^0C)	1.5	0.6	0.46
Melting point (^0C)	1415	937	1238

APPENDIX B

SOME PHYSICAL CONSTANTS

Boltzmann constant (k)	1.381×10^{-23} J/K
	or 8.617×10^{-5} eV/
Electronic charge (e)	1.602×10^{-19} C
Electron rest mass	9.109×10^{-31}kg
Electron volt (1 eV)	1.602×10^{-19} J
Permeability in vacuum (μ_0)	$4\pi \times 10^{-7}$ H/m
Permittivity in vacuum (ε_0)	8.854×10^{-12} F/m
Plank constant (h)	6.626×10^{-34} J/s
Velocity of light in vacuum (c)	2.998×10^{8} m/s

APPENDIX C

HOW TO PREPARE PROJECT REPORT ?

When we complete the project work we have to present it is an attractive way. Though there is no hard and fast rule for the preparation of reports it is customary to follow certain principles that are widely followed. When we write a research paper we are bound to follow the 'instruction to authors' of that particular journal. Here I will be giving you an idea about the preparation of a project report for your degree or postgraduate course.

While typing the project report, page setup is very important. Here are some points to be remembered for setting up pages in the report. The report should be neatly typed with a margin not less than 1¼ to 1½ inch on the left side and one inch on right, top and bottom. All pages should be numbered at the top/bottom centre. The line spacing should be 1½ or double. The fonts selected should be simple and legible with size '12'. The chapter heading, sub,headings,etc., should have sizes in proportion. Never use the same font size for chapter heading and sub headings. Throughout the report you should use the same font and font size for heading, sub heading and text.

The project report should contain the following sections.

1. Title page
2. Certificate from Guide/institution
3. Acknowledgement
4. Contents
5. Abstract
6. Introduction
7. Experimental/Theory part
8. Result and discussion - conclusion.
9. References

1. Title page

The tile of the project work should be concise but in formative enough to facilitate information retrieval, i.e. the title should be brief and yet convey the reader the nature of the contents of the report. Try to have a title containing less than 20 words.

Below the title after proper spacing write author's name. After this you have to write what and to whom you are going to submit the report. You may use the conventional sentence - *"The project report submitted to ________ University in partial fulfilment of the requirements for the award of the degree of Bachelor / Master of Science"* Arrange this sentence beautifully at the centre of the page. At the bottom of the page give the emblem and name of the institution and year of submission of the report. All these should be arranged aesthetically.

2. Certificate

On the second page give the certificate from you Guide stating that this is a bona fide record of the project work done by you.

3. Acknowledgement

On the third page you can acknowledge the people who have helped in successful completion of the project work. Don't forget to acknowledge your Guide and Head of the department. Otherwise you may be in trouble.

4. Contents page

The contents page should contain the title of all chapters with page numbers. It can also contain the headings *Abstract* and *References.* When there are large number of important subsections in each chapter you can modify the content page by incorporating them also. Then start the subsections 1.1, 1.2,etc., after a 'tab'. For example:

1. INTRODUCTION **Page No.**

1.1 ______

1.1.1______

2. EXPERIMENTAL TECHNIQUES/THEORY

2.1____

4. Abstract

The abstract should not have more than 200 words. It should convey the main points of the report, outline the results and conclusions and explain the significance of the result.

5. Introduction

In this section you have to give a brief introduction about the field of you work, essential principles, theories, etc. There can be subsections for this section. It should be numbered 1.1, 1.2,etc., and further subsections as 1.1.1, 1.1.2, etc. For example, if you are doing a project work on Physics of diode, you can give a brief introduction on semiconductors, pn junctions, characteristics, diode equations etc.

While writing the introduction you may be mentioning well known facts and results. The books or research papers you have used for carrying out the project work should be numbered consecutively in the order of occurrence in the text. There are several ways of giving the reference number in the text. You can give the number either as a superscript [22] or as a square bracket [22]. When you want to cite a reference repeatedly in the text, same reference number should be given everywhere. Full reference should be given in a list at the end of the report. When you prepare a large report, chapter-wise referencing system can also be adopted. Any how for college level project work chapter-wise referencing system is not essential.

Similarly all mathematical equations, tables and figures should be consecutively numbered in the order of occurrence in the text. For example, table 2 in chapter 1 can be given a number *Table 2.1*, a figure as *Fig. 2.1* and equation as (2.1) on the right margin (or after fixed number of tabs) of the same line of equation. The tables should be selfcontained and have a descriptive title. The pictures, graphs, etc., should be clear and should have a caption. On the graphs the axes should be marked clearly with proper units in the SI system. Follow the same system throughout the report.

6. Experimental or Theory section

This can be given as the second chapter. Here you can make a brief description of various experimental techniques related to your work. The essential theory behind the project work can be discussed in detail. There can be a number of sub sections or further sub section to this section.

Towards the end of this chapter the motivation or intention behind your project work should be given. Then discuss in detail your experimental setup, circuit diagram, design, theoretical formulation, etc.

7. Result and Discussion

In this section you have to give your observations either through tables or graphs. If you have done any standardization of your experimental procedure it should be given first. No need of both on the same data to convey a message. The tables should be self evident with proper units. In this section after standardization, if any, discuss in detail the way in which you have approached and analysed the problem, the way of calculation, the main results,etc., in a systematic order. If possible the final result can also be put in the form of a graph.

Make a brief discussion on the results you got. The limitations, possible errors, future suggestions for improvement, the accuracy of your result,etc., must be given. Do error analysis wherever possible. For the analysis of your result if you use any reported facts or equations the corresponding reference should be mentioned in the text and the details of reference should be cited at the end of the report. All symbols used should be explained when it appears for the first time itself.

In the conclusion section you can brief the results, the scope and application of your work. The limitations, further modifications, suggestions,etc., also can be include in this section.

8. References

While writing the report the references are numbered consecutively in the order of first appearance in the text and should be given in a separate list at the end of the report. Reference should be made to full list of authors rather than to first author followed by an abbreviation such as et al You can use et al in the text but not in the reference. In the event of multiple use of same reference the reference number should be the same. Do not try to cite unseen references.

For giving the list of references the following pattern can be followed.

For a book

S. Sankararaman, Introduction to Photonic Crystals, p. 122, Adhyayan Publishers & Distributors (2010).
i.e. Author, Book, page number, publisher, year.

For a paper in a journal

S. Sankararaman, *Appl. Phys. Lett.*, 67 (1995) 2939.
Author, Journal (Here Applied Physics Letters), Volume number, Year, Page number.

While using abbreviations for journals, follow those given

for Physical abstracts. It is a must to specify the author's initials and appropriate volume and page numbers in every case.

For internet web pages

Specify the exact address of the website.